Department of the Environment, Transport
and the Regions

Expert Panel on
Air Quality Standards

Lead

London: The Stationery Office

© Crown copyright 1998. Copyright in the typographical arrangement
and design vests in the Crown

Extracts of this publication may be made for non-commercial in-house use,
subject to the source being acknowledged

Applications for reproduction should be made in writing to The Copyright Unit,
Her Majesty's Stationery Office, St Clements House, 1-16 Colegate,
Norwich NR3 1BQ

Printed in Great Britain. Text printed on material containing 75% post-consumer waste.
Cover printed on material containing 75% post-consumer waste.

ISBN 0 11 753447 1

Previous reports by the Expert Panel on Air Quality Standards

1st report	Benzene	February 1994	ISBN 011 752859 5
2nd report	Ozone	May 1994	ISBN 011 752873 0
3rd report	1,3- Butadiene	December 1994	ISBN 011 753034 4
4th report	Carbon monoxide	December 1994	ISBN 011 753035 2
5th report	Sulphur dioxide	September 1995	ISBN 011 753135 9
6th report	Particles	November 1995	ISBN 011 753199 5
7th report	Nitrogen Dioxide	December 1996	ISBN 011 753352 1

United Kingdom air pollution information received from the automatic monitoring sites and forecasts may be accessed via the following media:

Freephone telephone helpline   0800 556677
CEEFAX   **pages 410-417**
TELETEXT   **page 106**
Internet   **http://www.environment.detr.gov.uk/airq/aqinfo.htm**

The cover photograph is reproduced by
kind permission of Professor Anthony Seaton

Contents

Page

iv	Dedication
v-vii	Terms of Reference and Membership of the Panel
viii	Acknowledgements
1	**Introduction**
3	**Sources of Lead**
5	**Exposure to Lead**
7	**Measurement and Monitoring of Lead**
11	**The Effects of Lead on Human Health**
13	**Justification of an Air Quality Standard for Lead**
15	**Recommendation for an Air Quality Standard for Lead**
16-18	Figures 1-3
19	Bibliography

DEDICATION

This monograph is dedicated to the memory of Dr Arthur Chamberlain, a leading authority on lead, who gave the Panel the benefit of his advice during the production of this report.

A Recommendation for a United Kingdom Air Quality Standard for Lead

Expert Panel on Air Quality Standards

The Expert Panel on Air Quality Standards (EPAQS) was set up by the Secretary of State for the Environment in 1991 following the undertaking, in the Environment White Paper 'This Common Inheritance' published in September 1990, to establish an expert panel to advise the Government on air quality standards. The terms of reference of the Panel are:

'To advise, as required, on the establishment and application of air quality standards in the United Kingdom, for purposes of developing policy on air pollution control and increasing public knowledge and understanding of air quality, taking account of the best available evidence of the effects of air pollution on human health and the wider environment, and of progressive development of the air quality monitoring network.'

This report is one in a series which deals with pollutants suggested to the Panel by the Department of the Environment, Transport and the Regions. Reports will be made on individual pollutants except where the Panel decide to deal with more than one because of the relationships between pollutants.

MEMBERSHIP OF THE PANEL

Chairman

Professor A Seaton CBE, MD, FRCP, FRCPE, FFOM
University of Aberdeen Medical School

Members

Professor J G Ayres BSc, MD, FRCP, FRSA
Birmingham Heartlands Hospital and University of Warwick

Dr P J Baxter MD, MSc, FRCP, FFOM
University of Cambridge

Professor P G J Burney MA, MD, FRCP, FFPHM
United Medical and Dental School's of Guy's and St Thomas' Hospitals, London

Professor R L Carter CBE, MA, DM, DSc, FRCPath, FFPM
Royal Marsden Hospital and University of Surrey

Dr J W Cherrie PhD, FIOH
University of Aberdeen and Institute of Occupational Medicine, Edinburgh

Dr A E Cockcroft MD, FRCP, DIH, FFOM
Occupational Health and Safety Unit, The Royal Free Hospital, London

Professor D N M Coggon MA, PhD, DM, FRCP, FFOM
MRC Environmental Epidemiology Unit, University of Southampton

Dr R G Derwent MA, PhD
Meteorological Office

Professor S T Holgate MD, DSc, FRCP, FRCPE
Faculty of Medicine, Southampton General Hospital

Observers

Dr H Jackson PhD
Health and Safety Executive

Dr P T C Harrison PhD, CBiol, MIBiol
Institute for Environment and Health, Leicester

Dr H Walton PhD
Department of Health

Mrs K Cameron MSc (until June 1997)
Department of the Environment, Transport and the Regions

Dr L Smith PhD (from June 1997)
Department of the Environment, Transport and the Regions

Secretariat

Dr M L Williams PhD
Department of the Environment, Transport and the Regions

Dr R L Maynard BSc, MB, BCh, MRCP, MRCPath, FIBiol
Department of Health

Dr S M Coster PhD
Department of the Environment, Transport and the Regions

Miss J Dixon MSc (until August 1997)
Department of the Environment, Transport and the Regions

Dr P R S Green PhD (from August 1997)
Department of the Environment, Transport and the Regions

Dr J B Greig MA, DPhil (until January 1997)
Department of Health

ACKNOWLEDGEMENTS

We thank the following individuals and organisations for their help.

AEA Technology, National Environmental Technology Centre, Culham, Abingdon, Oxfordshire OX14 3DB for their help with the analysis of United Kingdom air pollution data.

The late Dr Arthur Chamberlain for comments and advice.

Professor Stuart Pocock, London School of Hygiene and Tropical Medicine, for comments and advice.

Dr Marjorie Smith, Thomas Coram Research Unit, 27-28 Woburn Square, London, for comments and advice.

Dr Andrew Renwick, Clinical Pharmacology Group, University of Southampton, for comments and advice.

Introduction

1. Lead is second only to iron among the most widely used metals, having applications in the manufacture of batteries, pigments, alloys, plastics and ammunition. It has also been used widely in organic compounds as a petrol additive, although this application is now declining. It is no longer mined in the United Kingdom, but industrial workers may be exposed to it in smelting and refining operations, battery manufacture, scrap metal work, painting, soldering, ship repair and demolition, plumbing, manufacture of pottery and many other less common situations.

2. Lead can be absorbed into the body both through the lungs and through the stomach and intestines. Thus people may be at risk of absorbing it when exposed either in the air, dust, soil or as a contaminant in food and drink. In industrial situations there is a risk from inhaling lead-bearing dust or fumes from heated lead, and much of our knowledge of its harmful effects comes from study of such workers. Among the general public two sources of exposure are of particular importance; contamination of drinking water from lead pipes and contamination of the air from industrial sources and from combustion of leaded petrol. Lead in the air may not only be absorbed directly by the lungs but may also settle out and contribute to contamination of crops and of dust ingested inadvertently by children.

3. Lead has been known for centuries to be harmful to people working with it, and in particular has severe adverse effects on the blood, the nervous system and the kidneys. However, these clinical effects only occur as a consequence of high exposures and are relatively easily prevented. Of greater concern are the more subtle effects caused by lower exposures, such as may occur from the presence of lead in drinking water, paint and dust, and in the ambient air. The effects of lead on the intellectual development of children have been of especial concern. Children appear to be more susceptible to lead than adults, and may also absorb it to a greater extent when exposed.

4. In this report, the Panel discuss the main sources of lead exposure, including the relative contributions of lead in the air and lead in the diet, and the methods by which it is measured in the air. The Panel also consider the airborne concentrations recorded to date in the United Kingdom, ways in which lead is handled by the body, and its toxic effects on people. We then recommend an Air Quality Standard for the United Kingdom for lead.

Sources of Lead in Air

5. Lead occurs in the earth's crust and is released naturally through various processes including weathering of rocks, volcanic activity, and uptake and subsequent release from plants. Anthropogenic sources of lead stem from its removal from the earth's crust. It is released into the atmosphere through the mining and smelting of ores, the production, use, recycling and disposal of lead-containing products and the burning of fossil fuels. Industrial emissions and a large part of the vehicle emissions are in the form of particles of inorganic compounds of lead. Primary lead particles emitted from petrol vehicles are around 0.015 µm* in diameter and these aggregate to form larger particles with diameters of 0.1-1.0 µm; these particles can remain in the air for 7-24 days. Industrially emitted particles are around 0.1-5.0 µm in diameter depending on the process and the nature of the control devices employed.

6. The main sources of national airborne lead emissions in the United Kingdom for 1995 are shown in Table 1. Currently the dominant contribution is from petrol combustion at 1067 tonnes per annum. This results from the use of lead as an additive to increase the octane rating. In January 1986 the maximum permitted lead content of petrol was reduced from 0.40 to 0.15 grams per litre (g/l), and since then there has been a progressive increase in use of unleaded petrol to its current level of over 70% of the market. As a result, lead emissions into the air from petrol engined road vehicles in the United Kingdom have more than halved since 1987 (*Figure 1*). Under European Community legislation likely to be adopted in 1998, the marketing and sale of leaded petrol will be banned throughout the community from the year 2000, except in a limited number of specified circumstances. There are other sources of lead in air in addition to the lead emitted in motor exhausts. These include coal combustion, the production of non-ferrous metals and waste treatment and disposal.

---

* 1µm is one millionth of a metre.

Table 1. Estimated United Kingdom emissions of lead by emission source, 1995 (tonnes).

Source	Emissions (tonnes)	Percentage of Total *
Power Station Combustion	28	2
Commercial/Institutional/Residential Combustion	13	1
Industrial Combustion Plants and Processes		
Combustion Plant	44	3
Lead/Zinc/Copper	139	9
Other	31	2
Non-Combustion Processes		
Coke Ovens	1	<1
Iron and Steel Production	36	2
Chemical Processes	2	<1
Road Transport		
Petrol	1067	73
Diesel	1	<1
Waste Treatment and Disposal	105	7
Total	**1468**	**100**

Note:
*   Rounded to nearest 1%

Exposure to Lead

7. Direct human exposure to lead occurs not only through inhalation of particulate lead in ambient air, but also through ingestion of contaminated food, water and dust, and from occupational sources. In children and infants, ingestion of lead-containing dust following transfer from hand to the mouth is also important.

8. Food is the main source of lead intake for most people. Levels in food routinely monitored in the United Kingdom by the Ministry of Agriculture Fisheries and Food show that beverages, vegetables and milk are the main food groups containing lead. Lead may enter food through the deposition of dust and rain, containing the metal, on crops. In root crops, the contribution of deposited lead to the lead content of the edible portion of the plant is probably slight, but in leafy crops and cereals it may be more important. The amount of lead on food plants may be reduced through washing during food processing. Uptake of lead directly from the soil by food plants will also contribute to lead intake. However, only a small proportion of lead in soil is available for plant uptake and species differ widely in their uptake. Other ways in which lead may get into food include through food-processing activities, glazed ceramic dishes, lead crystal ware and, now rarely, lead solder in cans.

9. Drinking water may also be a source of lead exposure in the United Kingdom. The current lead standard for drinking water is 50 µg/l* at consumers' taps. In 1993 the World Health Organisation recommended a guideline value of 10 µg/l and this value has been included in the European Commission's revised Drinking Water Directive. Water leaving treatment works usually has concentrations of less than 10 µg/l. However, in areas where lead pipes, storage tanks or other fixtures and fittings are still in use, and the water is able to dissolve it, lead can leach into water resulting in occasional breaches of the current lead standard of 50 µg/l. Water supplies are treated to minimise this.

10. Soil and dust are major sources of exposure for children, who transfer dirt from their hands to their mouths, and dust can be the major component of their lead intake. In most rural and remote areas lead in soil is derived mainly from natural geological sources, but in urban and industrial areas soil lead is derived mainly

---

* 1 µg/l is one millionth of a gram of lead in 1 litre of water.

from man's activity. Additionally, the application of sewage sludge to agricultural land may increase the lead content of soil. Lead in outdoor dust is derived from the deposition of dust produced by lead-based paint, combustion of leaded petrol and from industrial sources such as lead smelters. The major source of lead in indoor dust is peeling and flaking of lead-based paints in older properties, although dust and soil from outdoors may also enter buildings.

11. Although normally only a small fraction of total lead intake occurs through inhalation, lead in air may also contribute to exposure indirectly, for example, through deposition in dust and on crops.

Measurement and Monitoring of Lead

12. The Department of the Environment, Transport and the Regions currently makes measurements of airborne lead at 24 sites across the United Kingdom in a variety of locations (*Figure 2*). Seven of the sites are located in urban areas, three are located at the kerbside of busy roads, eight sites are close to large lead works and six sites are in rural locations. *Table 2* shows the annual average lead concentrations recorded at each of these sites between 1980 and 1996. It can be seen that annual average concentrations in air have decreased at all sites over the 17 year period.

13. Ambient lead concentrations in the United Kingdom are generally measured using the "M-type" sampler, a technique in which air is drawn through a filter at a controlled flow rate allowing suspended particulate matter to be collected. The particulate matter is collected on filters which are changed weekly and are subsequently analysed for lead and, at some sites, other metals. The downward facing filter is held in a cylindrical hood, primarily for protection from wind and rain. This arrangement has been shown to collect particles which approximate to PM_{10} * when the wind speed is 6 metres per second, close to the average wind speed in the United Kingdom. At four of the rural sites (Chilton, Styrrup, Trebanos and Windermere) the "Harwell" sampler has been used. This sampler uses the same principle of operation as the "M-type" sampler in that air is drawn through a filter paper, though the design is somewhat different. The filters are changed monthly and are bulked for quarterly analysis for lead and other elements.

14. In recent years the analysis of lead in samples of particulate matter has been undertaken using a variety of spectrometric techniques, whereas, in earlier years other techniques such as X-ray fluorescence were used. Given these different analytical procedures, the accuracy of the data presented in *Table 2* is estimated to be between ± 5 and ± 15 %.

15. In the mid 1980s annual average concentrations of airborne lead at the kerbside of a busy road in West London were of the order of 1.4 $\mu g/m^3$**.

---

* Particulate matter less than 10 μm in aerodynamic diameter, (or more strictly, particles which pass through a size selective inlet with a 50% efficiency cut-off at 10 μm aerodynamic diameter).

** 1 $\mu g/m^3$ is one millionth of a gram of lead in every cubic metre of air.

Concentrations measured at urban sites located away from roads were broadly in the range 0.15-0.8 µg/m³.

16. *Figure 1* illustrates the decreases in both annual ambient lead concentrations (averaged for urban non-industrial sites) and emissions from petrol engined road vehicles after the reduction in the maximum permissible lead content of petrol from 0.40 to 0.15 grams per litre in 1986. The continuing steady decline in ambient concentrations in subsequent years reflects both the reduction of the lead content of leaded fuel and the increased consumer uptake of unleaded petrol. As a result, urban levels have reduced to the extent that the maximum values are now of the order of 0.2 µg/m³ even at the west London kerbside site at Cromwell Road.

17. Rural levels, as expected, are rather lower and currently range from about 0.009 to 0.038 µg/m³. In industrial areas in the vicinity of processes which emit lead, such as secondary non-ferrous metal smelters, levels can be higher than in urban areas where motor vehicle emissions are the main source of lead. In 1996 annual average concentrations at such sites ranged from about 0.117 to 0.882 µg/m³ (*Table 2* and *Figure 3*).

18. European Council Directive (82/884/EEC), which came into force in 1984, set a limit value for airborne lead concentrations of 2 µg/m³ as an annual average. Concentrations measured at the eight sampling sites around three industrial works in Walsall and Newcastle are used to monitor compliance with this Directive. This limit value was exceeded between 1985 and the end of 1989 at one industrially influenced site in Walsall. This site was in an area which was allowed temporary derogation from compliance with the Directive. Compliance was achieved in 1990, and there have been no further exceedences of the limit value.

19. In their 1987 publication the World Health Organisation set an air quality guideline for lead of between 0.5 and 1.0 µg/m³ as an annual average. The upper value of this guideline was not exceeded at any United Kingdom lead monitoring site in 1996. However two industrial sites in Walsall still exceed the lower limit of the guideline. A WHO working group has recently recommended the revision of this guideline to 0.5 µg/m³ as an annual average. The measurements from the existing monitoring network suggest that the proposed revised WHO guideline is unlikely to be exceeded in the urban environment, even at the kerbside of busy roads. However, this may not be the case in the vicinity of industrial plants which may be significant emitters of airborne lead.

Table 2   Annual average airborne lead concentrations in the United Kingdom, 1980-1996 ($\mu g/m^3$)

Site Name	1980	1981	1982	1983	1984	1985	1986	1987	1988	1989	1990	1991	1992	1993	1994	1995	1996
Kerbside																	
Cromwell Rd, London *				1.37	1.41	1.45	0.660	-	-	-	0.380	0.360	0.340	0.255	0.244	0.199	0.151
Cardiff	-	-	-	-	-	1.280	0.630	0.670	0.620	0.570	0.460	0.440	0.384	0.311	0.233	0.165	0.171
Manchester	-	-	-	-	-	2.040	0.810	0.810	0.760	0.640	0.510	0.460	0.339	0.305	0.123	0.133	0.118
Urban																	
Central London	0.640	0.580	0.630	0.470	0.520	0.480	0.270	0.280	0.300	0.220	-	0.120	0.099	0.078	0.085	0.060	0.074
Brent, London	0.770	0.710	0.890	0.990	-	0.640	0.300	0.290	0.320	-	0.220	0.200	0.174	0.147	0.144	-	0.148
Leeds	0.650	0.370	0.450	0.440	0.280	0.310	0.180	0.190	0.140	-	0.120	-	-	0.106	0.080	0.076	0.060
Motherwell, Strathclyde	0.260	0.230	0.300	0.240	0.180	0.260	0.190	0.180	-	-	0.200	0.160	0.050	0.086	0.023	0.050	0.030
Glasgow	0.460	0.330	0.240	0.420	0.190	0.270	0.120	0.180	0.130	0.140	0.095	0.092	0.093	0.090	0.039	0.051	0.052
Newcastle	-	-	-	-	-	0.180	0.130	0.150	0.110	0.110	0.070	0.070	0.067	0.070	0.027	0.025	0.035
North Tyneside	-	-	-	-	-	0.290	0.150	0.190	0.140	0.120	0.081	0.100	0.083	0.090	0.026	-	-
Rural																	
Cottered, Hertfordshire	-	-	-	-	0.130	0.130	0.077	0.098	0.076	0.075	0.041	0.045	0.044	0.036	0.019	0.020	0.025
North Petherton, Somerset	-	-	-	-	-	-	0.070	0.065	0.069	0.081	0.053	0.062	-	-	-	-	-
Eskdalemuir, Dunfries and Galloway	-	-	-	-	-	0.029	0.008	0.013	0.014	0.013	0.010	0.013	0.010	0.007	0.006	0.005	0.017
Chilton, Oxfordshire	0.110	0.056	0.066	0.065	0.086	0.090	0.033	0.006	0.052	0.048	0.038	0.038	0.027	0.022	0.025	0.025	0.025
Trebanos, West Glamorgan	0.082	0.080	0.078	0.092	0.098	0.081	0.039	0.048	0.055	0.064	0.043	0.063	0.060	0.037	0.032	0.039	-
Styrrup, Nottinghamshire	0.178	0.135	0.172	0.115	0.170	0.130	0.066	0.070	0.094	0.086	0.057	0.065	0.055	0.035	0.047	0.044	0.038
Windermere, Cumbria	0.047	0.039	0.047	0.045	0.048	0.035	0.024	0.020	0.023	0.021	0.015	0.020	0.008	0.008	0.012	0.013	0.009
Industrial																	
Walsall Metal Industries 1	-	-	-	-	-	-	0.930	1.370	1.160	0.880	0.470	0.570	0.540	0.570	0.500	0.700	0.579
Walsall Metal Industries 2	-	-	-	-	-	-	2.660	2.950	3.580	2.430	1.300	1.390	1.440	1.220	1.340	1.020	0.882
Walsall Metal Industries 3	-	-	-	-	-	-	0.760	0.730	-	0.640	-	0.340	0.410	-	-	-	-

Table 2 (continued)

Site Name	1980	1981	1982	1983	1984	1985	1986	1987	1988	1989	1990	1991	1992	1993	1994	1995	1996
Walsall Metal Industries 5	-	-	-	-	-	-	-	1.110	1.590	1.380	0.680	0.620	0.680	0.470	0.480	0.660	0.467
** Brookside 1	-	-	-	-	-	-	0.310	0.330	-	0.260	0.150	0.150	0.220	0.160	0.143	0.180	0.177
** Brookside 2	-	-	-	-	-	-	0.990	0.890	1.310	1.100	1.140	0.710	0.560	0.470	0.438	0.465	0.359
*** Elswick 1	-	-	-	-	-	-	-	1.650	1.350	0.530	0.370	0.660	0.340	0.190	0.335	0.480	0.546
*** Elswick 2	-	-	-	-	-	-	-	0.610	0.510	0.600	0.330	0.250	0.230	0.150	0.145	0.140	0.117
*** Elswick 6	-	-	-	-	-	-	-	-	-	0.780	0.450	0.460	0.300	0.190	0.200	0.190	0.215

Note:

*   the Cromwell Road site closed in September 1997, a new kerbside site at Marylebone Road, London became operational in February 1997.

**   sited at Walsall, West Midlands.

***   sited in Newcastle.

The Effects of Lead on Human Health

20. Lead is absorbed into the body both through the stomach and intestines after being taken in through the mouth, and through the lungs when breathed in from the air. Once absorbed, it spreads around the body and accumulates particularly in bone, teeth, skin and muscle. In these tissues it is relatively stable and released only over months or years. A small proportion, around 2%, is found in blood and it is this fraction that is biologically active and leads to harmful effects. Removal of lead takes place slowly through the kidneys, the concentration in the blood halving over the course of about several weeks in the absence of further uptake.

21. The toxic effects of lead are a consequence of its ability to inhibit the actions of certain enzymes and to damage chemicals in the nuclei of cells. In workers with high exposure a rare but serious manifestation of lead poisoning is acute brain damage, causing delirium and fits. Severe poisoning can also induce many other symptoms, and damage to organs such as the kidney can occur when concentrations in the blood exceed 100 µg/dl*. At somewhat lower concentrations, above about 80 µg/dl, colicky intestinal pains may be a feature. Above about 50 µg/dl anaemia can arise due to an inability to produce haemoglobin, the blood pigment that is responsible for carrying oxygen. Reversible effects on the kidneys and male reproductive organs have been described at blood concentrations greater than 40 µg/dl, as have effects on nerve functions in the limbs at concentrations above 30 µg/dl. Finally, above a level of 10 µg/dl studies of large groups of children have shown subtle evidence of changes in brain development, and this is also the lowest concentration at which biochemical evidence of interference with blood pigment synthesis has been described.

22. The most substantial evidence of effects of low levels of lead on health relates to effects on the central nervous system and, in particular, on the developing brain of children. Investigations have concentrated on average effects on populations. The end point most commonly measured has been the intelligence quotient (IQ), an index expressed in relation to the average (which is arbitrarily scored at 100) of the population as a whole. In any population the IQs of individuals are distributed around this figure so that those less intelligent than the average score less than 100, and those more intelligent score more than 100. Any substance that damages the

---

* 1µg/dl is one millionth of a gram of lead in every tenth of a litre of blood

brain might be expected to reduce the average IQ in the exposed population, an effect that would not be noticeable in people of average intelligence but that would increase the numbers of individuals with low intelligence and decrease the numbers of very intelligent people in that population.

23. Many studies have investigated the relationships between blood lead (or sometimes tooth lead) in children and IQ. While few of these studies have taken account of all other factors that might have been associated with lead exposure and have independently influenced intelligence (technically called confounders), the results of them all taken together suggest that there is an inverse relationship between blood lead and intelligence; that is, the higher the average blood lead concentration in a population, the lower that population's average IQ. In these studies it has not proved possible to show evidence of a threshold concentration in blood below which lead has no effects at a population level, although adverse effects in individuals have not yet been demonstrated below about 10 μg/dl. Some controversy surrounds these studies because of their inability to take account of all possible confounding factors, and it remains possible that the association between IQ and blood lead is not a causative one. Perhaps, for example, children of lower average intelligence are more likely to be exposed to lead because of their habits and environment. However, the studies have shown consistent results and are biologically plausible since lead is a known nervous system poison and can damage the brains of experimental animals. The Panel have therefore taken the prudent view that elevated lead concentrations in the blood do have the potential to cause damage to the developing brains of children.

24. The evidence from these studies of populations of young children suggests that the developing brain of a child from the time of birth up to the age of 5 years is at its most vulnerable and there may be a loss of up to about 2 IQ points on average for a rise in blood lead from 10 to 20 μg/dl. In view of this probable reduction in IQ and its effects on the numbers of people in the population with high and low intelligence mentioned in paragraph 22, there is a case for further action to reduce average blood lead levels in populations. As mentioned above, air lead concentrations are not the only determinants of blood lead in a population, since the major proportion of intake is derived from the diet. However, there is published evidence that reduction in air lead concentrations occurring over a period of increasing use of unleaded petrol has been associated with consistent reductions in the exposed population's blood lead concentrations, and the Panel have concluded that controls on airborne lead can reduce the risk to the health of the population through more than inhalation alone.

Justification of an Air Quality Standard for Lead

25. For the purposes of setting an Air Quality Standard for lead, the Panel have concluded that the critical health effect is on the intelligence of young children. A Standard which provides adequate protection against impairment of intelligence in children will also prevent the other toxic effects that have been linked with higher exposures.

26. Currently there is no convincing evidence of a threshold exposure to lead below which no effect on intelligence occurs. Therefore, in setting a standard the aim should be to identify a level at which any effect on intelligence is likely to be so small as to be negligible. The limit of accuracy for measurement of intelligence in individuals is 1 IQ point, and the Panel has taken the view that a lead concentration in the air that might cause an average fall in population IQ of 1 point should be regarded as unacceptable.

27. Epidemiological studies suggest that an increase in the concentration of lead in blood from 10 to 20 µg/dl is associated with an average reduction in population IQ of about 2 points. In the absence of evidence to the contrary, it is reasonable to assume that smaller increases in blood lead will have correspondingly smaller effects, and that therefore a rise in blood lead of about 5 µg/dl would be associated with a fall of about 1 point in population IQ.

28. The relationship between airborne concentrations of lead and the blood lead of children is complex. It is determined not only by absorption from the air that the child breathes but also by ingestion of lead in food and drink and deposited from the air onto surfaces as dust which the child may then transfer to the mouth. Available data indicate, however, that to increase blood lead concentrations by an average of 5 µg/dl, the airborne concentration of lead must be increased by about 1 µg/m^3.

29. Lead serves no useful biological function so ideally there would be no lead in air. If lead concentrations in air increased from zero to 1 µg/m^3, the above paragraphs indicate that health effects in children (shown as an average reduction in population IQ of about 1 point) would be detectable. Thus, the Panel have concluded that the standard could not be set at a concentration greater than 1 µg/m^3.

30. Even at 1 µg/m^3, there would be very little margin of safety. Consequently, we consider this figure should be reduced by a safety factor of 50% to take into account uncertainties in the relationship between blood lead concentrations and change in IQ and between air lead concentration and blood lead concentration. Bearing in mind that there may be some variation in the susceptibility of children to lead, we further consider that an additional 50% safety factor should be introduced to protect the most vulnerable. Thus the Panel recommends a concentration of lead in air of 0.25 µg/m^3 as an Air Quality Standard at which we believe any effects on health of children will be so small as to be undetectable and at which the vulnerable will be protected.

31. Lead at the concentrations found in the general environment in the United Kingdom does not pose a short term danger to health, but has an effect through long term exposure, since it is the total amount of lead that accumulates in the body that is important in determining adverse effects on the developing brain. The Panel, therefore, recommend that this Standard be applied as an annual average concentration.

32. When this Standard is exceeded, any risks to health are likely to be greater if the individual also has relatively high exposure to lead from other sources such as lead plumbing. In these circumstances, risk could be reduced not only by lowering airborne concentrations but also by controlling the other sources of exposure.

Recommendation for an Air Quality Standard for Lead

31. The Panel recommend an Air Quality Standard for lead in the United Kingdom of 0.25 $\mu g/m^3$ measured as an annual average.

32. This recommendation is intended to protect young children, the group regarded by the Panel as those most vulnerable to impairment of brain function. The Panel point out that action to protect such children should take account of all likely sources of exposure to lead. It is intended that techniques for monitoring the Standard should be consistent with the lead measurements made by the Department of the Environment, Transport and the Regions.

Figure 1 *UK annual estimated lead emissions from petrol engined road vehicles compared with mean annual ambient urban lead concentrations.*

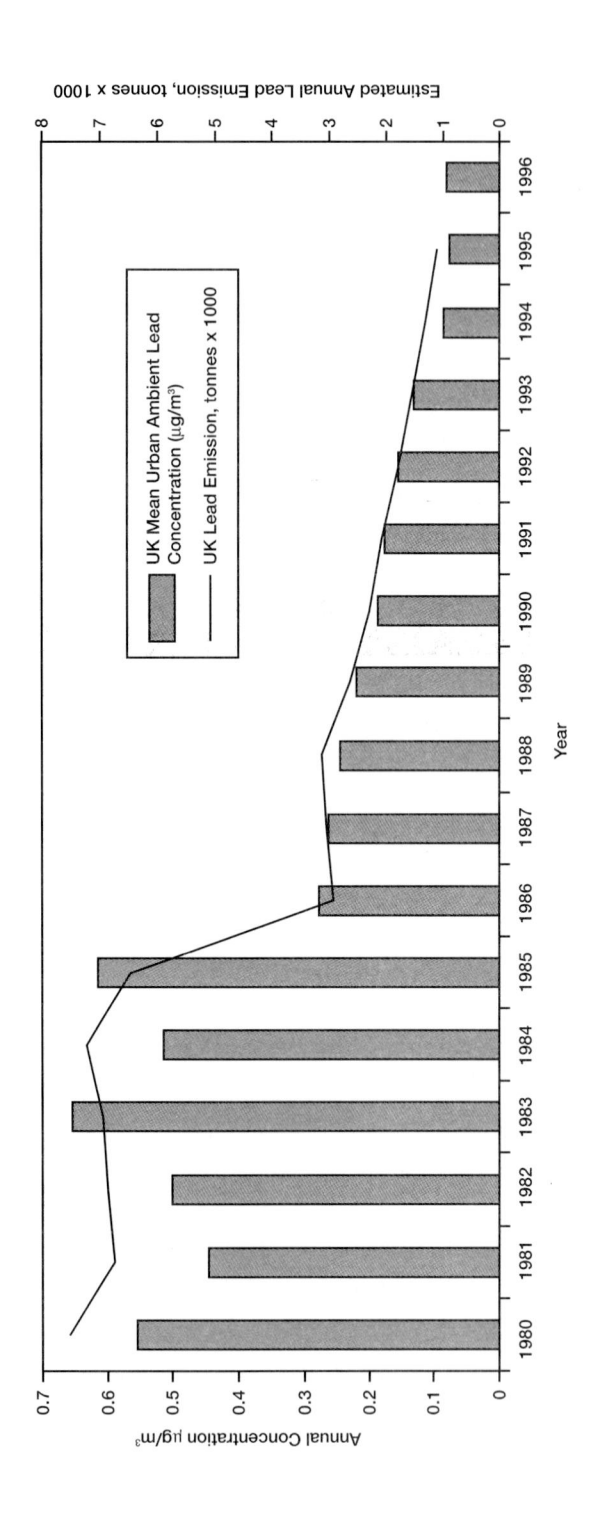

Figure 2 Department of the Environment, Transport and the Regions monitoring sites for lead as at the end of 1997.

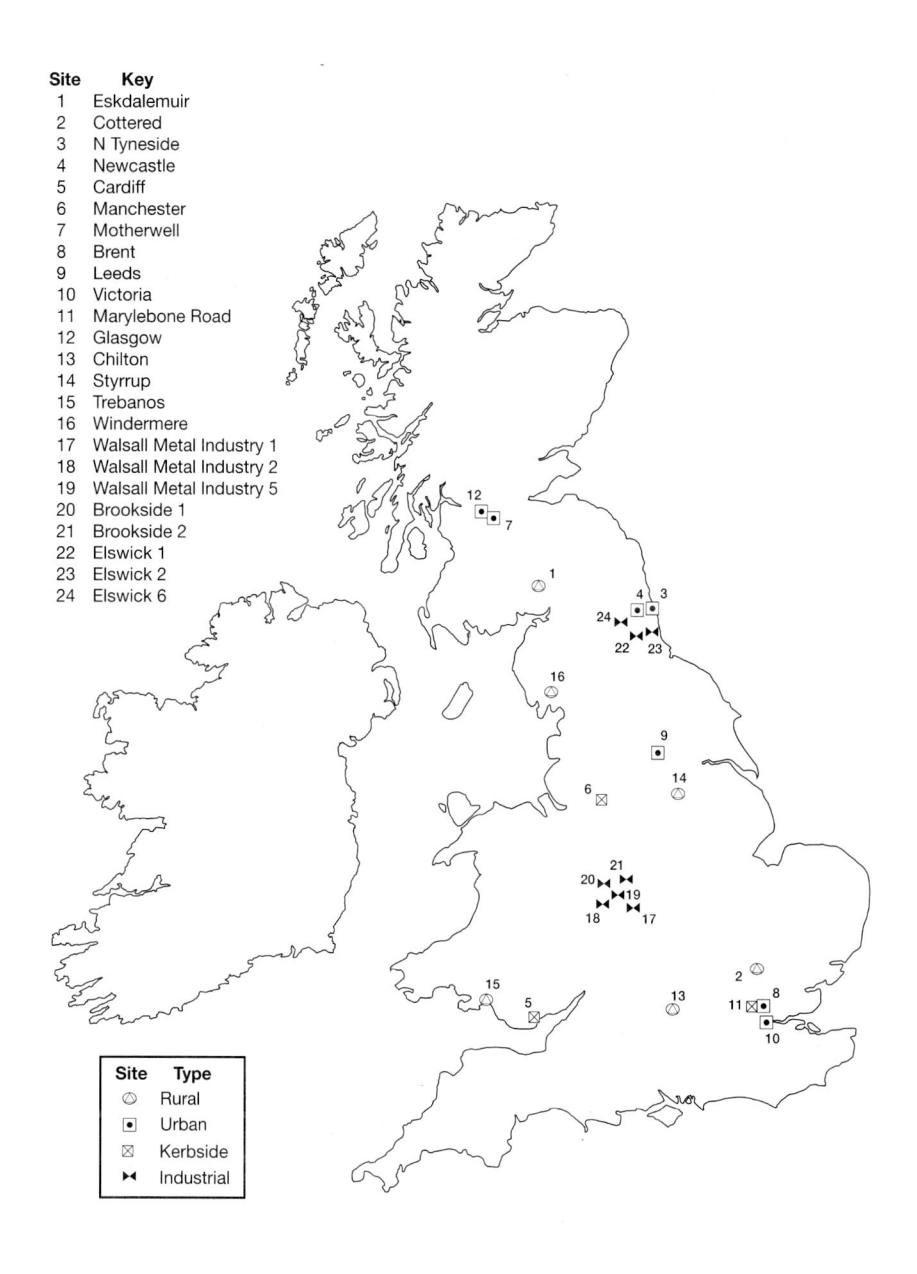

Site	Key
1	Eskdalemuir
2	Cottered
3	N Tyneside
4	Newcastle
5	Cardiff
6	Manchester
7	Motherwell
8	Brent
9	Leeds
10	Victoria
11	Marylebone Road
12	Glasgow
13	Chilton
14	Styrrup
15	Trebanos
16	Windermere
17	Walsall Metal Industry 1
18	Walsall Metal Industry 2
19	Walsall Metal Industry 5
20	Brookside 1
21	Brookside 2
22	Elswick 1
23	Elswick 2
24	Elswick 6

Site	Type
◯	Rural
▫	Urban
⊠	Kerbside
▶◀	Industrial

17

Figure 3 Annual average airborne lead concentrations at industrial monitoring sites 1986-1996.

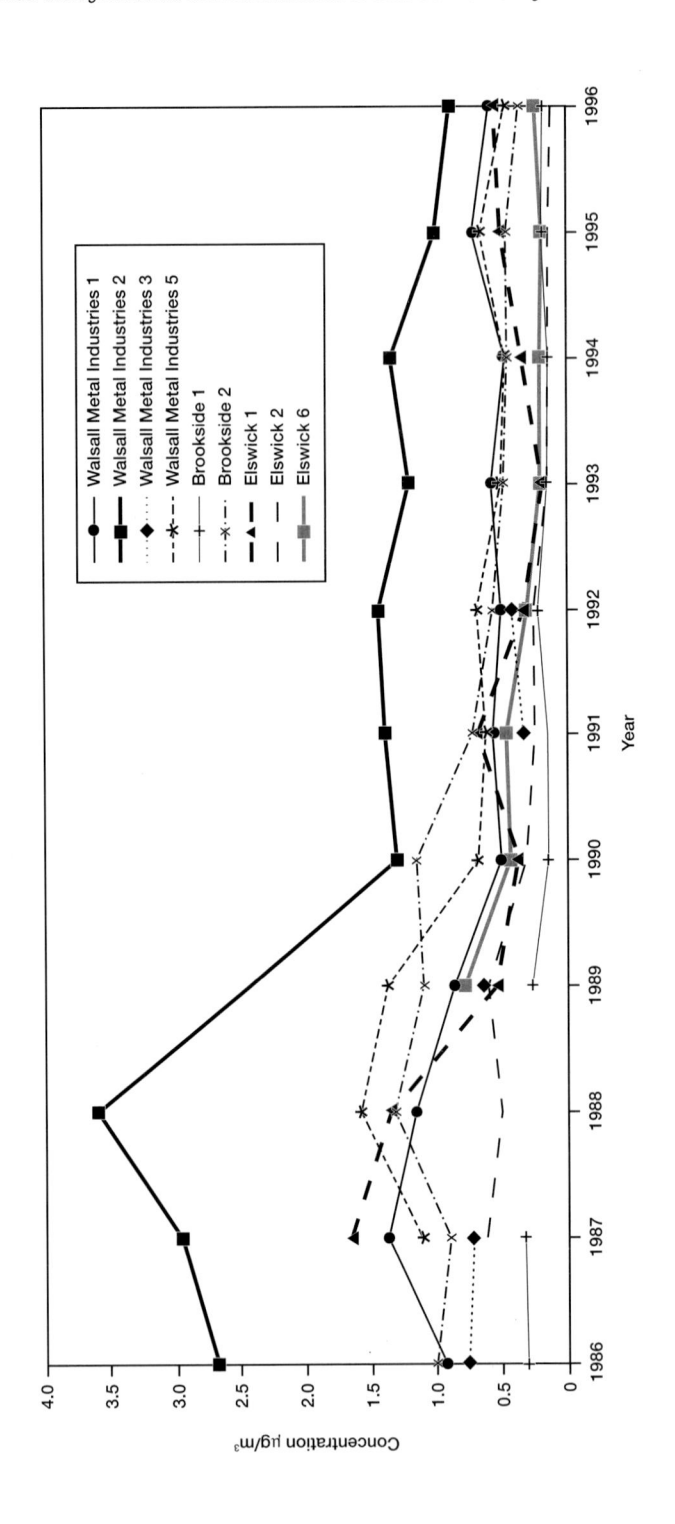

BIBLIOGRAPHY

Brunekreef B. The relationship between air lead and blood lead in children: a critical review. *Sci Tot Env* 1984; 38: 79-123.

Chamberlain AC, Heard MJ, Little P, Wiffen RD. The dispersion of lead from motor exhausts. *Phil Trans R Soc Lond A* 1979; 290: 577-589.

Davies DJA, Thornton I, Watt JM, Culbard EB, Harvey PG, Delves HT, Sherlock JC, Smart GA, Thomas JFA, Quinn MJ. Lead intake and blood lead in two-year-old UK urban children. *Sci Tot Env* 1990; 90; 13-29

Delves HT, Diaper SJ, Oppert S, Prescott-Clarke P, Periam J, Dong W, Colhoun H, Gompertz D. Blood lead concentrations in United Kingdom have fallen substantially since 1984. *BMJ* 1996; 313:883-884.

Department of the Environment, Transport and the Regions. Digest of Environmental Protection and Water Statistics, Number 19; HMSO: London; 1997.

Department of the Environment. Quality of Urban Air Review Group. Urban Air Quality in the United Kingdom. First report of the Quality of Urban Air Review Group. London, 1993.

Department of the Environment and the National Environmental Technology Centre. Report AEA/RAMP/20112002/002. Air Pollution in the UK: 1995; AEA Technology, Culham, Abingdon, Oxfordshire; 1996.

European Community Directive 82/884/EEC (1982) Directive on a limit value for lead in the air.

IPCS (1977) Environmental Health Criteria 3: Lead. Geneva, World Health Organisation, 160pp.

IPCS (1995) Environmental Health Criteria 165: Inorganic Lead. Geneva, World Health Organisation, 300pp.

Ministry of Agriculture Fisheries and Food (1994) 1991 Total Diet Study: Food Surveillance Information Sheet Number 34. Ministry of Agriculture Food and Fisheries, London.

Needleman HL, Gunnoe C, Leviton A, Reed R, Peresie H, Maher C, Barrett P. Deficits in psychologic and classroom performance of children with elevated dentine lead levels. *New Engl J Med* 1979; 300:689-695

Needleman HL. Correction: lead and cognitive performance in children. *New Engl J Med* 1994; 331:616-617

Pocock SJ, Smith M, Baghurst P. Environmental lead and children's intelligence: a systematic review of the epidemiological evidence. *BMJ* 1994; 309:1189-97

Tong S, Baghurst P, McMichael A, Sawyer M, Mudge J. Lifetime exposure to environmental lead and children's intelligence at 11-13 years: the Port Pirie cohort study. *BMJ* 1996; 312:1569-1575

World Health Organisation. *Air Quality Guidelines for Europe.* WHO Regional Publications, European Series No.23, Copenhagen, WHO Regional Office for Europe; 1987.

World Health Organisation. *Revised Air Quality Guidelines for Europe.* WHO Regional Publications, European Series, Copenhagen, WHO Regional Office for Europe; (*in press*).